FIGHTING FORCES IN THE AIR

F-16 FIGHTING FALCON

LYNN STONE

Rourke

Publishing LLC
Vero Beach, Florida 32964

www.rourkepublishing.com

PHOTO CREDITS: All photos courtesy of the U.S. Air Force

Title page: *The F-16 can fly combat missions in almost impossible weather conditions.*

Editor: Frank Sloan

Library of Congress Cataloging-in-Publication Data

Stone, Lynn M.
  F-16 Fighting Falcon / Lynn M. Stone.
      p. cm. -- (Fighting forces in the air)
  Includes bibliographical references and index.
  ISBN 1-59515-182-6 (hardcover)
  1. F-16 (Jet fighter plane) I. Title. II. Series: Stone, Lynn M. Fighting forces in the air.
  UG1242.F5S785 2004
  623.74'64--dc22
                                                    2004011745

Printed in the USA

CG/CG

TABLE OF CONTENTS

CHAPTER ONE

THE F-16 FIGHTING FALCON

The U.S. Air Force's F-16 Fighting Falcon was developed as a swift, lightweight, **multi-role** fighter aircraft. It has the **capability** and weaponry to both combat enemy aircraft and attack surface targets. F-16s have proven in combat that they can locate, reach, and destroy a variety of enemy targets.

F-16s were designed to operate in all kinds of weather and at night. F-16s are equipped with modern systems that can find targets even when the pilot can't see them.

An F-16C Fighting Falcon roars over ▶
northern Iraq.

▲

An F-16 prepares for a night attack against forces in Iraq.

In designing the F-16, aircraft engineers developed the world's best multi-role fighter aircraft. And it improves with age. F-16s are designed to accommodate new **avionics** as they become available. The Fighting Falcons have been continually upgraded with the latest advances in weapons, avionics, and communications systems.

Since production began on the F-16 in the late 1970s, more than 4,000 have been manufactured in more than 110 different versions.

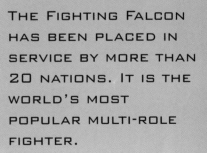

FACT FILE ★

THE FIGHTING FALCON HAS BEEN PLACED IN SERVICE BY MORE THAN 20 NATIONS. IT IS THE WORLD'S MOST POPULAR MULTI-ROLE FIGHTER.

F-16 Characteristics

Function: Multi-role fighter; air-to-air, air-to-ground attack aircraft

Builder: Lockheed Martin Aeronautics Company

Power Source: One Pratt and Whitney F100-PW-200/220/229 or General Electric F110-GE-100/129 (F-16C/D)

Thrust: 27,000 pounds (F-16C/D)

Length: 49 feet, 5 inches (14.8 meters)

Height: 16 feet (4.8 meters)

Wingspan: 32 feet, 8 inches (9.8 meters)

Speed: 1,500 miles per hour (2,400 kilometers per hour)

Ceiling: Above 50,000 feet (15,240 meters)

Maximum Takeoff Weight: 37,500 pounds (16,875 kilograms)

Range: More than 2,000 miles (3,240 kilometers)

Combat Range: approximately 575 miles (932 kilometers)

Crew: One (F-16C), one or two (F-16D)

Date Deployed: January, 1979

As advances in aircraft systems take place, many airplanes are changed from their original versions. If enough major changes occur, the Air Force may change the letter after the aircraft number. F-16Cs and Ds, for example, began to replace earlier F-16As and Bs in 1984.

The F-16C is designed for a single pilot. The two-seat F-16D can be flown by one or two pilots. The second seat is used by either a weapons system officer or a pilot instructor.

All active U.S. Air Force units and many Air National Guard and U.S. Air Force Reserve units with Fighting Falcons have only the F-16C and D models. The newest F-16s in production are E and F models.

The United States has sold the popular F-16 to several friendly nations. This Norwegian F-16 gulps fuel from a KC-135 Stratotanker in skies over Afghanistan. ▶

FLYING THE F-16

The F-16 is a highly **maneuverable** aircraft. It is compact and has a relatively lightweight **airframe**. An F-16 can climb quickly, peel into a screaming dive, and make tight turns. The U.S. Air Force's Thunderbird Air Demonstration Team flies F-16s.

▲
A missile-equipped F-16C flies over Washington, D. C., on a patrol mission.

▲

An F-16 banks into a turn.

F-16 pilots have excellent control of their machines partly because of the planes' novel "fly-by-wire" system. Before the F-16, modern fighter aircraft typically had cables and linkage controls. The F-16 introduced electric wires that relayed commands, replacing the cables and linkage controls.

In high gravity—**G-force**—maneuvers, an F-16 pilot uses a side stick controller instead of the traditional center-mounted stick.

Avionics in the F-16 include a computer-aided navigation system that directs steering information to the pilot. The plane has multi-band radios and an instrument landing system. It also has systems that detect the presence of either airborne or surface threats, such as missiles.

▲
The F-16C jets known as the Thunderbirds are the U.S. Air Force's precision flight team.

▲ *An Air Force technical sergeant attaches a Low-altitude Navigation and Targeting Infrared for Night Targeting pod to an F-16.*

The F-16 was the first operational military plane to receive GPS (Global Positioning System) capability. But F-16s are continually receiving upgrades of various systems. A new system can transmit images from a LANTIRN targeting **pod** to a video display in the cockpit. LANTIRN is an infrared navigation and targeting system. It permits low altitude and night attacks with a variety of weapons. Pods with LANTIRN electronics are mounted on the airplane's underbelly.

In 2002, F-16s began to receive the Joint Helmet-Mounted Cueing System (JHMCS). This system projects images onto the pilot's helmet visor. Any object the pilot sees—an enemy aircraft, for example—he or she can then target simply by looking at it through the visor and triggering the weapon release. Computers work out the details, even if the target is not in direct line of the weapon release.

◄ *A pair of F-16s rocket over northern Iraq.*

Under ideal weight and altitude conditions, the F-16 can fly at twice the speed of sound, or at **Mach** 2. The term Mach 1 indicates the speed of sound. The speed sound travels changes depending on low altitude (dense air) or high altitude (thin air). At 40,000 feet (12,192 m) above sea level, for example, the speed of sound is approximately 760 miles per hour (1,216 km/h).

Mach 2 flight is rarely necessary. F-16s usually fly at **subsonic** speeds, those below the speed of sound.

An F-16 is powered by one engine that produces power measured in pounds of **thrust**.

◄ *Powerful engine thrust sends an F-16C rushing down a runway in Turkey.*

CHAPTER THREE

FIREPOWER

Although the Fighting Falcon is relatively small—much smaller, for example, than the Phantom II fighters of the Vietnam War—it is a **lethal** combat machine. Pilots call their F-16s **vipers** because of their speed and deadly weapons. Series C/D F-16s are 49 feet, 5 inches (14.8 m) in length and have a 32-foot, 8-inch- (9.8-m-) wingspan. In comparison, the twin-engine Phantoms were 63 feet long (18.7 m) with a 38-foot, 5-inch- (11.8-m-) wingspan.

FACT FILE ★

AN F-16 CAN BE FITTED WITH A VARIETY OF WEAPONS, DEPENDING ON ITS MISSION. THE NEWEST F-16S HAVE 11 STATIONS TO WHICH MISSILES OR BOMBS MAY BE FITTED.

The F-4C Phantom II fighter of the 1960s was built in both Navy and Air Force versions.

THE LATEST F-16S CARRY AIR-TO-SURFACE HARM, MAVERICK, AND SHRIKE MISSILES. FOR ATTACKING SHIPS, AN F-16 CAN PACK PENGUIN AND HARPOON MISSILES.

The F-16 is the first fighter to carry a new family of "smart" weapons, the JDAM (Joint Direct Attack Munition). These new weapons can be fired accurately from fairly distant ranges in all weather conditions. F-16s also carry JSOWs, the new Joint Stand Off Weapons. JSOWs increase the distance from which they may be dropped with accuracy. That allows an F-16 to unleash the bomb from a greater distance, thus avoiding enemy ground fire.

▲ *Weapons hang under the wings of an F-16 on a combat mission in the Middle East.*

▲

An F-16C releases a 2,000-pound (907-kg) JDAM (Joint Direct Attack Munition).

In defense against other aircraft, the F-16 can launch the air-to-air AIM-7 Sparrow, the AIM-120 advanced medium-range air-to-air missile (AMRAAM), or the AIM-9 Sidewinder. In close-range combat, F-16s can fire the AIM-9X, IRIS-T, or Python series missiles. An F-16 can also fire 6,000 rounds of ammunition per minute through its 20mm six-barrel, wing-mounted Gatling gun.

F-16s are loaded with self-protection devices in addition to missiles. Jamming pods allow a pilot to broadcast signals that confuse enemy radar. F-16s can release flares or **chaff** to further confuse enemy defense systems.

CHAPTER FOUR

COMING OF AGE

The possibility for a quick-turning and economical multi-purpose fighter was raised in the early 1970s. The war experience in Vietnam had shown that the U.S. Air Force's big, extremely fast fighter jets could be out-maneuvered by smaller, slower, but more **agile** enemy fighters.

After three years of research and competition, General Dynamics' fighter jet—soon to be the F-16—was chosen for production. (Today the F-16 is manufactured by the Lockheed Martin Aeronautics Company, which merged with General Dynamics.)

▲

A Fighting Falcon fires an AGM-65D Maverick missile.

An F-16 bristles with missiles attached under its wings.

The F-16 design borrowed the best existing qualities and combined them with new ideas. The Fighting Falcon, for example, introduced improved cockpit instruments and pilot control. It introduced a reclined pilot seat for improved tolerance of G-forces. It brought in the fly-by-wire system and, for better visibility, the bubble canopy. It also provided the smaller size, lighter weight, performance, and relatively low purchase price that its designers had sought. The first combat-ready F-16s were sent to Hill Air Force Base, Utah, in January, 1979.

In 1991 F-16s flew combat missions in the Persian Gulf War (Operation Desert Storm). They flew more **sorties** there than any other aircraft. They attacked airfields, factories, missile sites, and other targets.

In the spring of 1999, Fighting Falcons conducted combat missions against Serbia during Operation Allied Force. F-16s destroyed radar sites, vehicles, tanks, buildings, and Serbian MiG fighter jets.

▲
One of the Air Force's F-16 Thunderbirds roars over the Nevada desert.

▲
The bubble canopy of the F-16 became the model that newer American jet fighters followed.

Fighting Falcons again flew in hostile skies in March and April, 2003, during the war against Iraq known as Operation Iraqi Freedom. One hundred and thirty-one F-16s flew nearly 4,000 sorties over Iraq.

CHAPTER FIVE

FLYING INTO THE FUTURE

The latest F-16s are F-16Es and Fs, first produced in the early 2000s. The F-16E/F series looks almost identical to the first F-16s, but it's a new airplane in the cockpit and beneath its skin. It carries more fuel and it has a bigger engine. Among other things, it has all-digital cockpit instruments, color screens, and a revolutionary electronic warfare system, the APG-80 Agile Beam Radar.

▲

A trio of F-16 "Vipers" streak across the skies of Iraq on a combat mission in March, 2003.

The F-16 will eventually be replaced in the U.S. Air Force by the Joint Strike Fighter. Meanwhile, F-16s will make up more than 50 percent of America's fighter force through 2010. Plans are for some F-16s to remain in the U.S. Air Force beyond 2020. Other nations will probably use F-16s into the 2030s. Lockheed Martin expects to continue building the fighters until at least 2008 and perhaps beyond.

▲

The Joint Strike Fighter shown here will eventually replace most of America's F-16 squadrons during a gradual changeover to newer aircraft.

Glossary

agile (AJ ul) — highly maneuverable; able to complete quick, graceful movement

airframe (AIR FRAYM) — the wings and shell, or body, of an airplane without its engines or weapons

avionics (AY vee ON iks) — the electronic systems and devices used in aviation

capability (KAY puh BIL it ee) — the ability to do something; being capable of

chaff (CHAFF) — small metal strips released into the air to confuse radar systems

G-force (GEE FORS) — the force of gravity

lethal (LEE thul) — deadly

Mach (MAWK) — a high speed expressed by a Mach number; Mach 1 is the speed of sound

maneuverable (muh NYUV ur ah bul) — capable of being changed in position for a specific purpose

multi-role (MUL tee ROLL) — capable of being used in more than one way

pod (PAWD) — a rounded compartment in which various electronic or other devices may be kept on an aircraft body

sorties (SORT eez) — more than one mission by one plane

subsonic (SUB son ik) — any speed below the speed of sound

thrust (THRUST) — the forward force of an object; the force produced by an aircraft engine

vipers (VY purz) — a group of highly venomous snakes, including the American rattlesnakes

INDEX

FURTHER READING

Chant, Christopher. *Fighters and Bombers: The World's Greatest Aircraft.* Chelsea House, 1999

Sweetman, Bill. *Supersonic Fighters: The F-16 Fighting Falcons.* Capstone, 2001

WEBSITES TO VISIT

http://www.fas.org/man/dod-101/sys/ac/f-16.htm

http://www.af.mil/factsheets

ABOUT THE AUTHOR

Lynn M. Stone is the author of more than 400 children's books. He is a talented natural history photographer as well. Lynn, a former teacher, travels worldwide to photograph wildlife in its natural habitat.